YOUR KNOWLEDGE HAS VALUE

- We will publish your bachelor's and
  master's thesis, essays and papers

- Your own eBook and book -
  sold worldwide in all relevant shops

- Earn money with each sale

Upload your text at www.GRIN.com
and publish for free

Bibliographic information published by the German National Library:

The German National Library lists this publication in the National Bibliography; detailed bibliographic data are available on the Internet at http://dnb.dnb.de .

Imprint:

Copyright © 2016 GRIN Verlag, Open Publishing GmbH
Print and binding: Books on Demand GmbH, Norderstedt Germany
ISBN: 9783668260641

This book at GRIN:

http://www.grin.com/en/e-book/322067/nanotechnology-as-treatment-for-lung-cancer

Daudi Nyangaresi

Nanotechnology as treatment for lung cancer

GRIN Publishing

GRIN - Your knowledge has value

Since its foundation in 1998, GRIN has specialized in publishing academic texts by students, college teachers and other academics as e-book and printed book. The website www.grin.com is an ideal platform for presenting term papers, final papers, scientific essays, dissertations and specialist books.

Visit us on the internet:

http://www.grin.com/

http://www.facebook.com/grincom

http://www.twitter.com/grin_com

Introduction

Studies relating to lung cancer treatment can save lives and make the lives of cancer patients more comfortable besides increasing the quality of life. According to Lardinois et al., (2003), there are three primary types of lung cancer; they include non-small cell, small cell, and carcinoid lung cancer. Statistically, non-small cell lung cancer, for example, Squamous cell carcinoma, adenocarcinoma, and large cell carcinoma account for the highest occurrences of all lung cancers to a tune of 85% [1]. It has been established that both environmental and genetic factors do contribute to lung cancer significantly. Continued exposure to carcinogens, ionizing radiation, and viral illnesses are known to bring about mutations of the DNA in the tissue covering the bronchial epithelium, particularly in the lungs. In addition to that, genetic anomalies in tumour suppressor gene inactivation and overactivity of growth generate oncogenes [2].

Currently, nearly 15% of individuals suffering from cancer live beyond five after diagnosis [3]. This indicates that diagnosis and treatment could be immensely augmented. Conventional medications treatments are surgery, radiation therapy [3-4], chemotherapy. Surgery is the main medical course of action mostly in the initial phase of lung cancers [4]. From the recent past, other new treatment designs have emerged. This study seeks to explore Nanomedicine because of its rampant use in lung cancer treatment in recent times. Nanomedicine simply connotes drug delivery systems manufactured from polymeric nanoparticles, liposomes or micelles. Nanomedicine technology is designed to target particularly the lungs.

2

Lung cancer poses challenges to humanity in the social, economic and health fronts. Economically, lunch cancer is taking an economic toll in western regions of the world than any other regions. This is the region that is worst hit with a rapidly increasingly economic burden associated with lung cancer unlike other forms of cancer. In 2011, the approximated overall healthcare costs for cancer treatment were $88.7 billion [5]. Lung cancer accounts for 20% of all of all healthcare costs related to cancer; amounting to about $17.74 billion [6]. In the European Union (EU), things are even worse; lung cancer is guesstimated to add to the biggest economic cost of all forms of cancer. It has been found to constitute up to 15% of all healthcare costs associated with cancer; translating to about $21.4 billion [7]. Lung cancer has been denoted as a high-cost condition, not only on the global scale but also on the level the patient [8].

Anatomically, lungs can be described as a pair air-filled organs situated on either side of the ribcage, commonly referred to as the thorax. The trachea delivers air into the lungs via the bronchi. The bronchi then split into form a pair of bronchioles [9]. The bronchioles head up straight to the alveoli absorption of air takes place. During the same time, carbon dioxide then from the blood lands in the alveoli it is breathed out. The lungs are well adapted to their functionally of expanding and contracting; they are covered by a thin layer of pleura.

It has been documented over and over again lung cancer causes more deaths in the US and the entire world in general than prostate cancer, breast cancer, and colon cancer altogether [11]. Since the mid-1980s, lung cancer has been reported as the common type of cancer globally with regards incidence. The incidence rates have

3

been rising over the years as well as mortalities related to lung cancer [11]. It is approximated that there are 1,350,000 new cases of lung cancer globally on an annual basis, accounting for about 12.4% of all new cancer diagnoses [11]. Regrettably, mortalities associated with cancer are estimated to be about 1,180,000 annually among other cancers accounting for 17.6% of overall cancer deaths [11]. Lung cancer also grades highest for probable cancer mortalities annually at 85,600 (28% of all cancer mortalities) for men and 71,340 (26% of all cancer mortalities) among women [11]. Whereas survival developments have been made the other types of cancers, nothing much has been achieved for lung cancer patients as attested by low survival rates of five years [11].

This paper seeks to explore the development of proper drug delivery systems (DDS). In doing this, the researcher will look at a wide range of various other types of cancers and the various DDS's used to create a better one on the basis of set criteria including type of nanocarrier, effective delivery, shrinkage of tumor, distribution of drug, biocompatibility/chemical reactivity and lastly possible side effects. Previously, DDS's took into consideration variety drug delivery systems including inhalable magnetic nanoparticles, hyaluronic acid–ceramide nanoparticles, personalized polypropylenimine (PPI) dendrimer, and stimuli-responsive clustered nanoparticles [12]–[15]. In the present study, the tactic of the DDS will encompass a hybrid constitution of numerous nanocarriers types that will be at the minimum of dual layers.

Methods: Developing a Nanotechnology Drug Delivery System

Overview of Nanotechnology Drug Delivery System

Nanoparticles that are used as vehicles to vehicles deliver drugs in the body are < 100 nm mostly in all dimensions. For therapeutic purposes, medicines can either be incorporated into the matrix of the particle or conjoined to the surface of the particles [16]. Drug delivery systems do regulate the fate of a drug incoming into the biological body system. Nanosystems with varying configurations and biological characteristics have been comprehensively studied for drug and gene delivery uses.

An ideal approach for realizing proficient drug delivery would be to develop judiciously nanosystems based on the comprehension of their interactions with the body, particularly the receiving cell population, and the changes in such cells with respect to disease progression. These observations compounded with observations made with regards to the mechanisms and location of drug action, molecular mechanisms, and pathophysiology of cancer can bring a better understanding of nanotechnology use in drug delivery systems.

It is imperative to study the barriers that hinder effective drug delivery. Some of these factors include of therapeutic carries in the living cells. Diminished drug efficacy may result from the volatility of the drugs once inside delivered into the cell, low availability as a result of various targeting or chemical interactions with the delivering carrier, changes in genetic composition of cell-surface receptors, sudden variations in in signaling conduits as cancer progresses as well as drug disintegration. For example, elevated levels of DNA methylation as cancer progress are known to lead to failure of multiple anti-neoplastic agents such as doxorubicin and cisplatin [16-17].

5

Proposed Nanotechnology Drug Delivery System

The proposed Nanotechnology Drug Delivery system will be created using a number of criteria as discussed below:

i)   *Type of Nanocarrier*

The utilization of nanotechnology in medicine has necessitated the creation of functionalized nanoparticles that can be utilized as carriers. Nanotechnology enables loading of the particles with drugs for delivery to specific locations in the body in a tightly controlled manner. As much as nanomedicine is quite new in the medical industry, some nanocarriers designed specifically for drug delivery have been in use for the past three decades. They include liposomes, dendrimers [16], quantum dots, viruses and virus-like nanoparticles [17] and more importantly to this study, polymeric nanoparticles. Of the various types afore-listed, this study narrows its scope to the use of nanogels. Nanogels can be defined as nanoscalar polymer systems that have a higher propensity to absorb water when put in contact with an aqueous environment.

Effective Delivery

As mentioned earlier, nanogels have a higher attraction to aqueous emulsions. Some other properties of interest include an impeccable colloidal firmness, non-reactivity to other blood substances as well as the internal aqueous system of the body. These properties make nanogels extremely suitable for integration of higher drug volumes and subsequent uptake and delivery of proteins, peptides as well as various biological compounds. Here, it is believed that the nanogels will synthesize dissimilar alternatives of poly(N-vinyl-pyrrolidone)-based nanogels with no cell toxicity thereby furthering the use of nanogels as an unparalleled drug delivery systems [18].

6

Shrinkage of Tumor

The use of nanogels in the treatment of targeted cancer through RNA interference (RNAi) and the suppressing EGFR by RNAi has yielded impeccable results. The utilization of the novel and an undoubtedly effectual method for the targeted delivery of delivery of drugs and DNA in the ovarian cancer cells has shown great results with respect to shrinkage of tumors. The approach is anchored on shell hydrogel which are nanogel carriers that present as an appropriate and adaptable structure for targeted drug and DNA administration. The shell nanogels are comprised primarily of alkylacrylamides that quickly integrated through multi-stage, free-radical instigated precipitation polymerization. During this occurrence, a permeable hydrogel core suitable for the entrapment of macromolecular drugs can be covered with a spongy hydrogel shell that shows the suitable chemoligation locations for the adhesion to pursuant ligands [19]

As mentioned earlier, there are certain properties of interest such good colloidal firmness, non-reactivity to other blood molecules and the internal aqueous systems of the body that make nanogels perform targeting effectively. In addition to this, nanogels have a higher affinity for water when put in contact with an aqueous environment. The affinity to water plays an integral function in tumor reduction or shrinkage in the target cells thereby increasing the functionality and efficiency of nanogel drug delivery system

Distribution of the Drug

Evaluation of the biodistribution following systemic administration nanoparticles is important. In the case of nanogels, the biodistribution centers on the primary tumor but also, it trickles down to disseminated metastasis [20].

7

Biocompatibility and Chemical Reactivity

Nanogels comprise of various biodegradable substances including natural and synthetic polymers, lipids as well as metals. The nanogels are taken up by cells more efficaciously than larger micromolecules and as such, could be utilized as proficient drug and DNA delivery systems.

Side Effects

Nanogels have minimal side effects is any, as such they present as the drug delivery system of choice. This is one of the primary reasons as to why they are of importance to this study. It is hoped that the study will benefit users of nanogels and more importantly promote its preference over other nanoparticles.

References

[1]   J. R. Molina, P. Yang, S. D. Cassivi, S. E. Schild, and A. A. Adjei, "Non-Small Cell Lung Cancer: Epidemiology, Risk Factors, Treatment, and Survivorship," *Mayo Clin. Proc.*, vol. 83, no. 5, pp. 584–594, May 2008.

[2]   K. M. Fong, Y. Sekido, A. F. Gazdar, and J. D. Minna, "Lung cancer. 9: Molecular biology of lung cancer: clinical implications," *Thorax*, vol. 58, no. 10, pp. 892–900, Oct. 2003.

[3]   W.-H. Lee, C.-Y. Loo, D. Traini, and P. M. Young, "Inhalation of nanoparticle-based drug for lung cancer treatment: Advantages and challenges," *Asian J. Pharm. Sci.*, vol. 10, no. 6, pp. 481–489, Dec. 2015.

[4]   T. Sadhukha, T. S. Wiedmann, and J. Panyam, "Inhalable magnetic nanoparticles for targeted hyperthermia in lung cancer therapy," *Biomaterials*, vol. 34, no. 21, pp. 5163–5171, Jul. 2013.

[5]   "Economic Impact of Cancer." [Online]. Available: http://www.cancer.org/cancer/cancerbasics/economic-impact-of-cancer. [Accessed: 04-Apr-2016].

[6]   P. J. Goodwin and F. A. Shepherd, "Economic issues in lung cancer: a review.," *J. Clin. Oncol.*, vol. 16, no. 12, pp. 3900–3912, Dec. 1998.

[7]   R. Luengo-Fernandez, J. Leal, A. Gray, and R. Sullivan, "Economic burden of cancer across the European Union: a population-based cost analysis," *Lancet Oncol.*, vol. 14, no. 12, pp. 1165–1174, Nov. 2013.

[8]   C. E. M. Desch, B. E. M. Hillner, and T. J. M. Smith, "Economic considerations in the care of lung cancer patients. : Current Opinion in Oncology."

9

[Online]. Available: http://journals.lww.com/co-oncology/abstract/1996/03000/economic_considerations_in_the_care_of_lung_can cer.9.aspx. [Accessed: 05-Apr-2016].

[9]   "The Lungs (Human Anatomy): Picture, Function, Definition, Conditions," *WebMD*. [Online]. Available: http://www.webmd.com/lung/picture-of-the-lungs. [Accessed: 04-Apr-2016].

[10]   "Anatomy and physiology of the lung - Canadian Cancer Society." [Online]. Available: http://www.cancer.ca/en/cancer-information/cancer-type/lung/anatomy-and-physiology/?region=mb. [Accessed: 04-Apr-2016].

[11]   C. S. Dela Cruz, L. T. Tanoue, and R. A. Matthay, "Lung Cancer: Epidemiology, Etiology, and Prevention," *Clin. Chest Med.*, vol. 32, no. 4, pp. 605–644, Dec. 2011.

[12]   J.-E. Chang, H.-J. Cho, E. Yi, D.-D. Kim, and S. Jheon, "Hypocrellin B and paclitaxel-encapsulated hyaluronic acid–ceramide nanoparticles for targeted photodynamic therapy in lung cancer," *J. Photochem. Photobiol. B*, vol. 158, pp. 113–121, May 2016.

[13]   V. Shah, O. Taratula, O. B. Garbuzenko, O. R. Taratula, L. Rodriguez-Rodriguez, and T. Minko, "Targeted Nanomedicine for Suppression of CD44 and Simultaneous Cell Death Induction in Ovarian Cancer: An Optimal Delivery of siRNA and Anticancer Drug," *Clin. Cancer Res.*, vol. 19, no. 22, pp. 6193–6204, Nov. 2013.

[14]   H.-J. Li, J.-Z. Du, X.-J. Du, C.-F. Xu, C.-Y. Sun, H.-X. Wang, Z.-T. Cao,

X.-Z. Yang, Y.-H. Zhu, S. Nie, and J. Wang, "Stimuli-responsive clustered

nanoparticles for improved tumor penetration and therapeutic efficacy," *Proc.*

Natl. Acad. Sci., p. 201522080, Mar. 2016.

[15]   A. Kulkarni, P. Rao, S. Natarajan, A. Goldman, V. S. Sabbisetti, Y.

Khater, N. Korimerla, V. Chandrasekar, R. A. Mashelkar, and S. Sengupta,

"Reporter nanoparticle that monitors its anticancer efficacy in real time," *Proc.*

Natl. Acad. Sci., p. 201603455, Mar. 2016.

[16] Suri, S. S., Fenniri, H., & Singh, B. (2007). Nanotechnology-based drug

delivery systems. *Journal of Occupational Medicine and Toxicology, 2*(1), 16.

[17] Rigogliuso, S., Sabatino, M. A., Adamo, G., Grimaldi, N., Dispenza, C., &

Ghersi, G. (2012). Polymeric nanogels: Nanocarriers for drug delivery

application. *CHEMICAL ENGINEERING, 27.*

 [18] Dickerson, E. B., Blackburn, W. H., Smith, M. H., Kapa, L. B., Lyon, L. A.,

& McDonald, J. F. (2010). Chemosensitization of cancer cells by siRNA using

targeted nanogel delivery. *Bmc Cancer, 10*(1), 1.

[19] Kumar, D., Saini, N., Pandit, V., & Ali, S. (2012). An insight to pullulan: a

biopolymer in pharmaceutical approaches. *International Journal of Basic and*

Applied Sciences, 1(3), 202-219.

[20] Huang, X., Zhuang, L., Cao, Y., Gao, Q., Han, Z., Tang, D., ... & Wang, S.

(2008). Biodistribution and kinetics of the novel selective oncolytic adenovirus M1

after systemic administration. *Molecular cancer therapeutics, 7*(6), 1624-1632.